Marsden Manson

Geological and Solar Climates

Their Causes and Variations. A thesis

Marsden Manson

Geological and Solar Climates
Their Causes and Variations. A thesis

ISBN/EAN: 9783337138981

Printed in Europe, USA, Canada, Australia, Japan

Cover: Foto ©berggeist007 / pixelio.de

More available books at **www.hansebooks.com**

Geological and Solar Climates

Their Causes and Variations.

———————

A THESIS.

BY

MARSDEN MANSON, C. E.

Geology and Physics:
UNIVERSITY OF CALIFORNIA,
May, 1893.

SAN FRANCISCO:
George Spaulding & Co., Printers,
414 Clay Street.

ERRATA.

Preface, next line to last, for *waiver* read *waver*.

Page 5, foot-note, 7th line from bottom, for *Hetvetic* read *Helvetic*.

Page 14, in foot-note, 6th line from top, for *Zeno-graphic* read *Zenographical*.

Page 17, last line, for *area* read *era*.

Page 20, 14th line from top, for *wherever* read *whenever*.

Page 23, 3d line from bottom, for *area* read *era*.

Page 23, 4th line from bottom, for *merging* read *emerging*.

Page 41, 15th line from top, after *necessary for the,* insert, *removal of glacial conditions, and for the*

Page 44, the second paragraph should read: The trapping process not being a function of the orbital distance, nor of the actual amount of heat received, but of the composition of the atmosphere, this rise, etc.

Page 47th, 6th line from top, insert " at end of paragraph, after out.

Ib, 6th line from bottom, for * substitute †.

PREFACE.

The worshipers of truth are delving in every hamlet—many have before them the daily burdens of life, from which they can snatch but a few hours each day to give to their chosen faith.

Every now and then one comes forward with some skillfully carved jewel which he has wrought into shape to deck his shrine. Sometimes it is only a little piece merely good for inlaying the walls, yet it fits well in its place and strengthens the faith of other workers. Again it is the great keystone for some massive arch whose other stones were laid in bygone times. Yet again it is a mighty truth that will not fit in the great building at all until the wrong work be torn down, and then it forms the base for one of the steadfast and everlasting towers. So pure must be the faith of those who bow at the hallowed shrines of truth that they would tear down these shrines rather than let them stand upon, or even harbor error.

The writer gives in this little book a keystone which he knows will not fit in the present building unless some errors be torn out. Those whose faith is true will not waiver nor come grudgingly to the work of rebuilding.

GEOLOGICAL AND SOLAR CLIMATES,

THEIR CAUSES AND VARIATIONS.

THE CAUSE OF THE ICE AGE.

" The most important problem in terrestrial physics * * and the one which will ultimately prove the most far reaching in its consequences, is: What are the physical causes which led to the Glacial Epoch and to all those great secular changes of climate which are known to have taken place during Geological Ages?" (Dr. Croll, Climate and Cosmology.)

"An attentive study of the physical Geography of the earth and its influences on Climate, together with a judicious application of the simplest physical theories, will enable us to gain by and by a better knowledge of Geological climates." (Prof. A. Woeikof, Nature, March 2, 1882, p. 426.)

Since Agassiz announced* the past existence of an age during which ice covered temperate and tropical land areas, *the cause* of this wonderful phenomenon has been a problem of profound interest. Upon the correct solution of it hinges also the cause of Geological climates.

---

*In 1821 Venetz called attention to the once greater extension of Glaciers; and in 1824 Prof. Esmark made similar observations as to the Glaciers of Norway. Phil. Mag., Vol. XXVII, p. 321.

As early as 1821 a prize of 300 livres was offered by the Helvitic Soc. of Nat. Sci. for the collection of facts regarding the increase or decrease of the extension of glaciers in the Alps. Tillock's Phil. Mag., Vol. LVII, p 307.

But to Agassiz belongs the honor of having first pointed out the existence of *the* Ice Age wnen *all* glaciers were vastly more extended than at present.

So great has been the interest attaching to this subject, that more study has been devoted to it during the
past fifty years than perhaps to any other in Geology;
hardly a leading scientific magazine runs through a
year's numbers without one or more articles upon it;
and no Geological Society is without zealous students of
glacial phenomena. Some have become so absorbed in
the subject that, led by the recurrence of certain slight
astronomical influences, they recognize a glacial period
for slight and widely scattered evidences of possible
early local glaciation, forgetful of the fact that an era of
frigid climate could not intervene between two eras of
tropical climates without the intervention of eras of temperate climates.

The evidences establishing the reality of the Ice Age*
during the Quaternary period are now beyond dispute.
It is difficult, however, to establish by geological evidence
the synchronal glaciation of all the continental areas
known to have been heavily glaciated. This difficulty
arises from the fact that the identity of various strata
has to be established by fossils of varying conditions
and characters; it is also rare that the same geologist
has visited and compared the evidence upon more than
two continents, thus eliminating probable errors from
unequal subaerial denudation and exposure in the different zones of present climates and upon different continents. Again, the proof of the contemporaneous existence
of corresponding strata upon different continents in the
same latitude is sometimes attempted by a comparison

---

* The writer prefers the nomenclature of Dr. Geikie and others using the
term Ice Age rather than Glacial Epoch, or Period. The duration of this
age was not recorded in the same manner and terms as either previous or
succeeding ages; this is due to the inactivity, or even absolute suspension
of the great forces, heat and moisture, over continental areas during this
age; under estimates of its duration are thus liable to be made.

of land fauna and flora, with marine fauna and flora, or even by more complex comparisons. Fossil plant life is by far more reliable than animal life for comparative purposes.

Another misleading factor is found in the interpretation of the great trans-continental lines of terminal moraines into the absolute limits of glaciation. Considering the great lapse of time since the removal of glacial conditions in temperate and tropical latitudes, it is more than probable that the existing unobliterated evidences by no means mark the extreme limits of a lighter and more extended glaciation whose traces have been destroyed, but which can justly be interpolated between the existing very marked traces of enormous glacial extension during Quaternary times. It is not impossible, nor entirely improbable, that early local glaciation did not occur during the early part of the Cenozoic Era, or even earlier, but the data upon which to establish the occurrence of such early local glaciation are both meagre and obscure. Should the evidences of such early local glaciation be developed beyond dispute, they will in no way interfere with the interpretation to be given, but they will strongly corroborate certain portions of this interpretation. So far as the author has been able to examine such evidence, it has been found to be between strata containing fossil life of a torrid character, with no evidences of a gradual merging into a temperate climate above and below it, as in Quaternary glaciation.*

Before entering further into this discussion, it may not be out of place to briefly review the principal theories advanced to account for the Ice Age. It will be seen that physicists and astronomers have vied with geolo-

---

* This question is so important and has so broad a bearing that it will be reverted to later under the heading *Palæozoic Glaciation.*

gists in the diligence of the search for the cause of this age, and their minds have been as fertile in the number of causes assigned as the true one. Not one of all the causes suggested has been sustained by argument without a flaw in the reasoning, and no demonstration has been made which has carried conviction to the scientific world.

It would not be instructive to attempt to review all of the theories which have been urged. The tendency to ascribe remote inadequate or obscure causes, rather than to interpret facts and phenomena in accordance with known laws, is apparent in many. Some writers have ascribed causes resting only upon hypotheses beyond the range of either analysis or investigation; such hypotheses can only stand in the absence or failure of all other assignable causes. Therefore the leading causes only will be briefly mentioned.

In a recent monograph on the subject, the following are given:*

1.   A decrease in the original heat of the globe.

2.   Changes in the elevation of land, and consequent variations in the distribution of land and water.

3.   Changes in the obliquity of the axis of the earth.

4.   A period of greater moisture in the atmosphere.

5.   Variations in the amount of heat radiated by the sun.

---

* Transactions of the Technical Society of the Pacific Coast, Sept., 1891, Vol. VIII. See also The Climate Controversy, S. V. Wood, Jr., Geol. Mag., 1876 and 1883. Climate and Time, Climate and Cosmology, Croll. Island Life, Alfred Russell Wallace, F. R. S., etc. Philosophical Magazine, May, 1864. British Association Reports, part 2, p. 11. Proceedings Royal Soc., vol. xxviii, p. 15. Quart. Jour. Geological Soc., Feb., 1878. Nature, July 4, 1878. Trans. Geological Soc., Glasgow, Feb. 22, 1877. The Ice Age in North America, Dr. Fred. G. Wright, Appendix by Warren Upham. The Cause of an Ice Age, Sir Robert Ball, F. R. S., etc. Révolutions de la Mer. Déluges Périodique, Alphonse Joseph Adhémar.

6. A variation in the heat absorbing power of the sun's atmosphere.

7. Variations in the temperature of space.

8. A coincidence of an Aphelion winter with a period of maximum eccentricity of the earth's orbit.

9. A combination of 8 and 2.

10. The views of Sir Robert Ball, LL. D., etc., as expressed in his recent work, *The Cause of an Ice Age.* *

The first of these theories is universally admitted, and taught in even elementary works on Physical Geography, but it fails to account for all the phenomena accompanying the Ice Age, or to account for the disappearance of that age, and, so far as the author is aware, has not been presented in such form as to satisfactorily account for geological and present climates in rigid conformity with the facts and known laws. Nor has it been presented in such form as to account for that era of geological climates known as the Ice Age; moreover, it fails to account for the disappearance of that Age.

The second has been proved to be a local and correlated phenomenon, but cannot be accepted as a *cause*, since glaciation did not solely depend in the same latitudes upon elevation above sea level.

As to the third, whilst slight changes in obliquity have occurred, and must continue to occur, the results are too slight and the distribution of glacial phenomena is too general to warrant the acceptance of such change as a prime cause.

The fourth is a necessary consequence of the first, but, like the first, fails when the crucial test of accounting for the disappearance of the continental Ice Sheets is applied.

---

* See also *The Date of the Last Glacial Epoch.* Gen. Drayson, R. A. Science, Nov. 25, 1892.

The fifth, sixth and seventh theories are mere hypotheses, unsupported by either demonstration or observed facts.

The eighth has been presented to the scientific world through the labors and researches of that eminent geologist and physicist, Dr. James Croll, in his various articles in leading scientific magazines, and lastly, in his grand contributions to the subject under discussion, "Climate and Time" and "Climate and Cosmology."

The ninth has been maintained by one of the greatest English naturalists, Mr. Alfred Russell Wallace. He combines the theory of Dr. Croll with that of Sir Charles Lyell, and very ably presents his views in "Island Life."

The tenth is a presentation by Dr. Ball, F. R. S., etc., of an interesting demonstration, to the effect that 63 per cent. of solar heat reaches either hemisphere during its summer exposure, and the remaining 37 per cent. during winter exposure. Nothing is added to the Physical Theory of Dr. Croll, nor does the demonstration in any way remove the serious objections which have been urged against Dr. Croll's views.

The strongest support that has been given to any of the above theories is made by the arguments and deductions of Dr. Croll and Mr. Wallace; yet they have failed to produce conviction, for, in a recent work on Geology, the author, after reviewing the various theories as to the cause of the Glacial Period, uses this expression: "This seems to be by far the most probable yet presented."*

This opinion is directly given upon only one—the ninth; but its terms are such that it embraces all. If

---

* Elements of Geology, Le Conte, 2d Edition, page 578.

the ninth is " by far the most probable," it would be difficult to fix the degree of probability or improbability of the others.

The only explanation which can be accepted is one which will admit of definite proof, and will satisfy all the conditions, and not require the distortion of known facts, by forcibly fitting them into arbitrary molds. It must start from universally admitted premises, and in rigid consonance with known laws, correctly interpret the grand eras of climate which have marked the geological history of our globe, and further, it must point out and fully elucidate wherein and why the present climates of our globe differ so radically from those vast secular variations recorded by fossil life—aye, more, it must be so general as to be of universal force and applicable to other members of the solar system constituted as our globe.

In the brief review just made of the principal theories urged by various scientists as causes producing the Ice Age, it was remarked of the first that it was universally admitted as true, and even taught in elementary works on Physical Geography, but that it failed to account for all the facts developed by the Ice Age. This first theory was a decrease in the original heat of the globe, the truth of which is established by a mass of indisputable geological evidence.

The present conditions are so radically different from any of the eras of climate known to have existed, that the explanation of this range of secular changes becomes the grandest problem in terrestrial physics, and has an important bearing in the solution of existing conditions upon the other planets.

It is universally admitted that this original heat has been so lost that it is no longer a factor in the surface temperature of the earth, and that solar energy is now the controlling source of heat.

There can then be no mistaking the first nor the present condition of the earth as regards its exposure to the only two sources of heat—(1) solar and stellar * heat, and (2) resident, internal, or earth heat.   There can therefore be no error as to the main features of the problem.

There must have been two marked eras of climatic control—(A) a past era, during which both sources were active; (B) and the present era, in which the greater exterior source only remains, the local and lesser source having been practically exhausted.

Or, in other words, we have, *first*, a heated globe having resident in its mass a finite quantity of heat, undergoing loss and exposed to an exterior source of heat and light, which source may be either constant or decreasing in its energy, but so slowly that it may be considered sensibly constant during the eras under consideration; *second*, the same globe deprived of its heat to such an extent that a crust of non-conducting material has formed, the outer surface of which is exposed only to solar heat, and whose climates are entirely controlled thereby.   The objects in view being to explain (1) the peculiar uniformity of climates prior to the exhaustion of the first source, and (2) the occurrence of an age of general glaciation in all latitudes prior to the establishment of the sole control of the exterior source; (3) the reasons of the differences between heat distribution during geological and present climates.   Such explanations to be in strict conformity with admitted facts and

---

* Stellar heat having the same function as solar heat, and being sensibly a constant of unknown amount but much less than solar heat, need not be separately considered in further discussion of the question.

known laws, and without omitting the one nor distorting the other.

To be explicit we will state that the prime objects are to demonstrate—

1. That in the passage of the earth from an era during which its climates have been controlled by internal heat into an era during which its climates are controlled by solar heat, eras of uniform climates must have been passed through during which isotherms were independent of latitude.

2. That before climates could have passed under solar control that an age must occur during which continental areas must be glaciated; and that this stupendous phenomenon, occurring before solar climatic control, was also independent of latitude.

3. That the direct *cause* of the Ice Age was a combination of the remarkable properties, in relation to heat and cold, possessed by the various forms of water. As *vapor*, in the form of fogs and clouds it prevented the loss or receipt of heat by radiation; as *water*, by reason of its high specific heat, it retained to the last moment the effective remnant of earth heat; as *ice*, it assumed a solid form, storing the maximum amount of cold.

4. To point out in a general way the fallacies of previous attempts to explain geological and present climates.

The problem will be given in a general proposition, which is capable of demonstration in perfect accord with known laws.

[This demonstration was first given by the Author in September, 1891, and is reproduced here slightly modified and extended from Vol. VIII of the Transactions of the *Technical Society of the Pacific Coast.*]

THE GENERAL PROPOSITION.*

GIVEN.—*A heated globe, constituted and circumstanced as the earth, and whose surface temperatures, by reason of internal heat, are above the boiling point of water, to prove that before its surface temperatures can pass under the control of the solar heat* (1) *that climatic changes must be independent of latitude, and* (2) *that the continental areas†️ must be glaciated.*

It will be observed that the surface temperatures of a globe thus situated are entirely controlled by its own internal or earth heat; for between such surface and any external source, a dense cloud of vapor must exist. The fact that direct or radiant heat rays cannot pass through dense fogs and clouds is well known;‡ therefore, a globe

---

* The proposition here stated is applicable to any planet. It is probable that Mars, Venus and Mercury have passed through periods corresponding to our Ice Age; and that Jupiter, Saturn, Uranus and Neptune have not reached theirs. A study of Jupiter in this connection is particularly instructive. Phenomena are presented which are easily explained by the theory under discussion. See Zenographic Fragments, London, 1891. Also the author's views in *Circulation of the Atmosphere of Planets*. Trans. Technical Society of the Pacific Coast, Vol. XI, No. 4, pp. 127–143.

† There were circumstances protecting certain areas, such as the "Unglaciated Area," in the basin of the Yellowstone River, in North America. Here vast and continuous lava overflows in Wyoming, Idaho, Oregon and Washington, liberated earth heat; which heat, borne easterly by the general circulation of the atmosphere, caused the precipitation upon the "Unglaciated Area" to be warm rains instead of snow. To this region the animals of the tertiary period retreated as glacial conditions surrounded them. Here they were protected, and perpetuated their species, and in these regions vast quantities of their remains are found. *Mining and Scientific Press*, Feb. 14, 1892.

The easterly projection of the unglaciated area is opposite the corresponding projection in the lava overflow. How this simple explanation has escaped the researches of geologists is not known.

‡ Maury, *Physical Geography of the Sea*, 6th Edition, p. 212, *et seq.* Croll, *Climate and Time*, p. 60, *et seq.* Also, Climate and Cosmology, p. 51.

thus situated can neither give off, nor receive radiant heat. The peculiar function of solar heat during the existence of appreciable quantities of earth heat was to warm the upper regions of the atmosphere and the outer surface of the clouds exposed to its power, thus partly replacing the heat lost by radiation into space, and causing the store of earth heat to last longer.

By the conditions of the problem presented, we thus have a globe having resident in its mass a finite quantity of heat exposed to loss only by means of the gradual expansion of water into vapor, and the exposure of this vapor to loss of heat by radiation from its upper surface into space. This vapor would then condense, and as rain, snow or hail, descend all, or part of the way to the earth, receive another increment of heat, and ascend as before. A slow process, but exhaustive in time.

Thus the property of water to assume three forms, each of which possesses remarkable qualities with regard to heat and cold, afforded the only means for exhausting the earth heat. As vapor, it possesses the property of storing more heat than any other known substance;* as snow or ice it possesses the property of storing more cold than any other known substance. The function of solar heat, until the exhaustion of earth heat by this process, was simply conservative; it merely warmed the upper layers of the atmosphere, through whose dense vapor its heat rays could not pass. Clouds being more translucent than transcalent, light rays reached the planetary surface prior to heat rays.

The earth may thus be regarded as having been surrounded by a series of spheroidal isothermal shells of mean temperatures. The one next the surface repre-

---

*Except hydrogen.

sented a mean temperature of $212° + t°$ Far.; t being positive, and proportioned to the greater pressure of the heavier atmosphere existing.   Above this isothermal shell were others representing mean temperatures of 90°, 60°, 32°, Zero, etc., to—x° Far., the extreme cold of interplanetary space.   Between the two spheroidal iostherms of 32° and—x° Far., was one which had a mean temperature of 32°—y°, and equally exposed to both sources of heat.

That the spheroidal isotherm of 32° Far. was within the sphere of influence of earth heat, is proven by the formation of snow or ice at that temperature, both being the resultant of vapor expanded and raised by earth heat to that height as a minimum.   Moreover, vapor would have reached that height as a minimum were solar and stellar heat suspended for a definite period, and the earth absolutely exposed to loss by radiation with no partial return of heat from exterior sources.

It therefore follows that the isotherm equally heated by both exterior and interior sources was colder than 32° Far. or below that temperature at which snow and ice form.   (It is well known that solar energy cannot maintain a temperature as high as 32° Far. except in the lower regions of the atmosphere.)

The isothermal shells nearest the earth were spheroidal in shape, and by reason of the conditions their surfaces were practically parallel with that of the earth; those most remote from the earth, by reason of solar influences, protruded at the equator and flattened at the poles, so as to be slightly more oblate than the earth; they were sensibly parallel with the spheroidal isotherm now marked by the "snow line."   Hence at the equator the direct action of the sun was first felt and established.

As the earth heat was a finite quantity exposed to loss, it was in time exhausted. As this loss proceeded, these spheroidal isothermal shells of mean temperatures shrunk in upon the earth, and their contact with its surface marked the zones of corresponding climates prevailing during the dual source of heat. Since these isotherms were independent of equatorial or polar exposure to solar energy their contacts with the planetary surface established climates independent of equatorial or polar position, or in other words of latitude; and not until those, whose distance from the surface mainly depended upon solar energy, shrunk to the surface could climates ranged in latitudinal zones be established. As the climates established by the contact of the isotherms inside of $32°-y°$ Far. were independent of direct solar heat, they varied from the climates established by solar heat alone; hence the marked difference between climates antedating and succeeding the Ice Age. The isotherms preceding this age were dependent almost entirely upon elevation above sea level, fractures and conductivity of the earth's crust; those succeeding it are dependent upon proximity to the equator, elevation above sea level, and the distribution of heat by ocean currents.

At the expiration of a period of time T., the earth lost sufficient heat to cause the isothermal shell of $90°$ Far. to shrink to the surface except at fractures, and a particularly uniform, moist, and highly torrid climate was established, and types of life developed, culminating in the Carboniferous Age.

The crust cooled sufficiently to permit the demarkation of the continental areas, but the cooling did not proceed to that point which upheaved the massive mountain ranges, nor greatly depressed the ocean areas. Therefore, an era of low, flat continents, and shallow hot

2

seas followed.   The life of that period abundantly shows this condition from one pole to the other, and the prevailing temperature is distinctly recorded in the fossil life of the Palæozoic and Mesozoic Eras.

Light rays reached the surface prior to this time, as evidenced by the development of visual organs in animal life.

The greater part of the vapors and gases existing previously in the atmosphere were condensed, and existed upon the surface; the vapors as highly heated oceans, and the gases in various combinations of the mineral and life kingdoms.   Now, in the oceans thus formed and further enlarged, there was stored up a vast quantity of the original earth heat, by reason of the *high specific heat of water*, from which it was not exhausted until the last moment; and in this process of exhaustion, it must have maintained the cloud shield, shutting out solar heat until this the last remnant of effective earth heat was exhausted.   Not only this, the oceans thus formed had a mean temperature of $90° + z°$ Far., z being a positive increment due to the heat received from the bottoms and sides of the ocean.   Not until the bottoms of the oceans were subjected to a degree of cold approximating that to which the continental areas were exposed could the crust be cooled uniformly and reach that degree of uniform thickness and stability suitable to the safety and comfort of the human race.*

---

* Moreover, in cooling the subjection of one pole to glacial and the other to temperate or sub-tropical conditions, as argued by Dr. Croll, would have subjected our planet to very peculiar "cooling strains," as they are termed by foundrymen.   Whereas the slow and uniform cooling, as herein described, is productive of maximum thickness, strength and uniformity of crust; and, as will be explained later, this crust was finally shrunk in upon the interior mass by being subjected to the maximum degree of cold to which it can be exposed during the existence of the sun as a source of heat.

At the expiration of the period of time, T', the spheroidal isothermal shell having a mean temperature of 60° Far., similarly shrunk to the surface of the earth, and a corresponding uniformly temperate climate was established. –

The further cooling of the crust caused its shrinkage, and a consequent greater upheaval of those areas most exposed to loss of heat, the continents. This further shrinkage caused the strata formed during the previous eras to be upheaved and fractured, and the lines of demarkation between oceans and continents were thus more strongly accentuated.

The life developed in the interim evidences an approach to that of the present temperate zones, and its wide distribution demonstrates the complete control of the climates of the globe by internal heat. The isothermal lines were entirely at variance with those established by solar heat, therefore the functions of solar heat remained conservative of those operating on the surface during this period also.

The extreme and uniform distribution of fur or hair-covered animals and of the deciduous and coniferous trees of the Cenozoic era mark further the control of a source of heat more uniformly distributed than solar heat could possibly be. For reasons previously given, this isotherm also reached continental areas earlier than ocean areas. When the mean temperature of the land was 60° the tepid oceans must have had a mean temperature of $60°+y°$ Far., y, like z, being positive, and due to increments of earth heat received from the bottom.

At the expiration of this period T', or at some time, $T' \pm a$, the isothermal shell of 32° Far. shrunk so as to reach the more elevated portions of the continental areas, and thus established a snow line independent of the influences now establishing and maintaining such snow line. The resulting glaciation was controlled by the same general laws that now exist, only the distribution of heat being independent of latitude, and mainly dependent upon altitude above sea level, glaciation of present tropical and temperate latitudes was as certain to occur as in polar regions. The moment a snowflake reached the earth which the waning earth heat was unable to melt, the Ice Age was inaugurated; and the conditions were such as to favor its extension until the exhaustion of the store of heat beneath the oceans and resident in them, by reason of the high specific heat of water. It will be noted here that wherever, in obedience to the expansive force of this waning earth heat, a particle of water was vaporized and made the last round of its circulation, it returned to the earth in that form which stored the maximum degree of cold, or, in other words, in that form which required the maximum amount of solar heat to change.

From the moment that snow began to accumulate, every remaining vestige of earth heat was available for producing those conditions favorable to glaciation, namely, warm seas, dense fogs and cold continental areas; and every unit of solar energy reaching the upper regions of the atmosphere was available for maintaining those favorable conditions.* Glaciation under these conditions would be cumulative until the oceans, exhausted of their heat and lessened in area,

---

* The prime objection which is urged against all previous theories is their inadequacy. We here have a perfectly adequate cause—resident earth heat to supply evaporation and shut out solar energy, which energy can only act the part of a conservator of the glacial conditions until the exhaustion of earth heat, when its power can be spent in melting glacial ice, and in gradually establishing the present conditions.

were no longer able to supply the moisture necessary to completely shroud the earth from direct solar heat.

At the expiration of the time T″, the isothermal shell, having a mean temperature of 32° Far., shrunk in upon the globe, and the oceans were exhausted of their store of heat and their bottoms brought in contact with water having a mean temperature of 31° Far., a temperature approximating that of the ocean depths at present, and of ice in masses.

The isothermal shell 32° Far. was a spheroid circumscribing the earth. In shrinking to the earth its intersections with the surface were controlled by the elevation of the surface above sea level, and by the local escape of earth heat; elevated equatorial or temperate areas were therefore as much exposed to glaciation as polar lands. (For maximum depth of glaciation see page 32.) By reason of the high specific heat of water, this isotherm also reached continental areas prior to reaching ocean areas.

The crust beneath the ocean, having been protected from loss of heat by the superincumbent water, shrunk to its final shape subsequent to that portion forming continental areas. The ocean bottoms in thus shrinking approximately to their present shape must have been fractured, as continental areas had previously been. In this way very considerable increments of earth heat were set free after glaciation had commenced. This process, which is entirely in consonance with known laws, would result in increasing the depth of glaciation, or even in re-establishing it after partial recedence.

There would also result a complicated series of crust movements as the continents were relieved of pressure by the melting of the ice caps, and the ocean bottoms subjected to increased pressure by the restoration of water to the oceans.*

Thus the same forces which, even before the eras we have been considering, must have built up upon the surface of the globe mineral forms of surpassing beauty, only to be destroyed and ground down to give place to vegetable and animal forms of wonderful development—these same forces were called upon to well nigh obliterate every living individual of both kingdoms. The efficiency of their work is attested in every zone of life from the equator to the poles.

The exhaustion of the residuum of earth heat in the oceans and beneath them could only have been accomplished by the same means as before, and this exhaustion resulted in the preservation of those conditions most favorable to glaciation.   When by the chilling of the oceans to about 31° Far. and by the glaciation of continental areas, the air was cleared of obscuring clouds and fogs, the wonderfully uniform series of climates was at an end.

With the dominion of solar heat there dawned upon our planet an era of climatic zones whose lines sensibly follow parallels of latitude; then also began seasons of spring, summer, autumn and winter, with the varying changes of the earth's annual round.

The climatic changes during the control of earth heat,

---

* It will again be noted that the isotherms inside of 32°—y° Far. were maintained by earth heat, and therefore independent of equatorial or polar exposure to solar heat.   Consequently their intersections were upon different lines from those isotherms exterior to 32°—y° Far., which latter were mainly dependent upon solar heat.   It will also be observed that solar heat, when it reaches the lower, denser regions of the atmosphere, is trapped and therefore capable of establishing and maintaining higher temperatures than in the upper atmosphere.

and within the range of geological research extended over eras:

1. An era of torrid heat.
2. An era of tropical heat.
3. An era of temperate heat.
4. An era of glacial cold.

Each merged gradually into the others, but each recorded its period of existence in unmistakable terms, all shrouded from the direct action of solar heat, and all evidencing by the life produced, the stifling, smothered character of the climate.

That solar heat was shut out from the surface of the earth during the Ice Age is geologically recorded in the glaciation of the North Temperate Zone over continental areas, where solar energy has removed glacial cold and established in its stead a mean annual temperature of 40° Far., and in the torrid zone it has removed glacial cold and established a mean annual temperature of 76° Far., where snow never falls.

Consequently, in a heated globe, constituted and circumstanced as the earth, exposed to two sources of heat, internal heat and solar heat, before its climates or surface temperature can pass under the control of solar heat climatic changes must be independent of latitude and the continental areas must be glaciated.

A General Comparison of the Demonstration with the Facts of Geology.

The dawn of the Archæan Era found the earth a heated globe merging from an unrecorded and unfathomable area of greater heat. The crystalline character of the earliest rocks demonstrates the high temperature which prevailed upon the surface at that time. Such being the temperature of the surface, it is beyond question that the existence of uncombined water upon it was an impossibility, and as vapor it could only shroud the earth in dense clouds. The earth heat was as effectually shut *in* from loss by radiation as was solar heat shut *out* from reaching the surface.

As this finite amount of earth heat could only escape by doing *work* in the expanding of water to vapor, vast eras of time must elapse before the work done could exhaust the available heat. The process of exhaustion was further retarded by two causes: 1st, the heating of the outer layers of the atmosphere by solar heat; and 2d, the low conductivity of the strata of the earth itself; consequently the climates of the earth, until the final exhaustion of earth heat, being controlled by a uniformly distributed supply, were of remarkable uniformity. The denudations, faults and fractures of its crust set free additional increments of heat but slowly, so that the torrid, tropical and temperate eras were longer than the frigid era.

During the existence of sensible quantities of earth heat the oceans must have been heated from the bottom, and cooled at the surface by evaporation. The evaporation from the total ocean surface under such conditions would give rise to much more extensive cloud formations than at present. Indeed, the record of temperatures and character of life are such as to warrant—nay, even force—the conclusion that the whole earth was one vast hothouse, from which solar heat was shut out, and throughout which a uniform temperature was prevalent from pole to pole.

Solar heat does not penetrate the thinnest cloud; even a fog through which the form of the sun is distinctly

visible shuts out nearly all direct solar heat.* The failure
in the past to recognize the climatic influence which the
factor earth heat was able to produce, and the endeavor
to ascribe to solar energy the climatic conditions existing during the activity of earth heat, has caused all the
mystery and error of attempts to explain the climatic
phenomena prior to and during the Ice Age.

Once realize the peculiar influence and domination of
earth heat, and these mysteries and errors fade, and the
whole system of preglacial and glacial climates becomes
simple.

The function of solar heat during the activity of earth
heat could be none other than conservative of the latter;
such function it is now performing for the great planets,
Jupiter and Saturn, and probably Uranus and Neptune,
whose surfaces are shrouded from our view by clouds.†

---

* The truth of this fact is easily established by either observation or
experiment. At the close of a hot day should a slight cloudiness super-
vene, the loss of heat by radiation from the surface is checked; the air at
the surface, and the surface, remain at the same temperature, and nature's
delicate differential thermometer—the deposition or non-deposition of
dew—records the non-transcalency of clouds, in terms worthy of consideration.

Again, let two delicate thermometers be exposed, one to the air temperature and the other in addition to direct solar rays; the latter will mark the
increased temperature due to such exposure. Upon the intervention of a
cloud, or even a jet of steam, both instruments will mark the same temperature.

See Physical Geography of the Sea. Maury, 6th ed., p. 212, *et seq.*
Climate and Time, Croll, p. 60, *et seq.*
Climate and Cosmology, Croll, p. 51.

† Astronomers agree that there must exist upon Jupiter a high degree of
heat, and yet no refinement of thermometric determinations can detect
any more heat from the Jovian surface than should be reflected from the
sun.—See *Young's General Astronomy*, page 353; also *History of Astronomy
During the Nineteenth Century.* Clerke, pp. 335-338.

Climatic Facts Established by Fossil Life.

It would be impossible, in the limits to which it is necessary to restrict this paper, to review the vast array of facts which could be brought forward to demonstrate the perfectly uniform, torrid character of the climates of the globe during the Palæozoic Era.

From the 81st degree of north latitude through every range of present climates to the confines of the south frigid zone, the life systems attest the stifling hothouse character of the climate. The species of plant life and animal life, whether of land or marine forms, varied less from the torrid to the frigid zones than corresponding species upon different continents in the same zone do now. Nowhere below the Permian deposits can fossil life be recognized that does not belong to an ultra-tropical type. Such uniformity of temperature is impossible under solar control, and hence can only belong to a climate controlled by earth heat.

In reviewing the temperatures recorded by the fossil life of the Palæozoic Era, the fact becomes apparent that nowhere upon the surface of the globe during that era were there any zones of temperature. The whole surface was subjected to one universal torrid climate—the life developed was uniform in its general character from the Arctic to the Antarctic circle. Under no possible conditions could such uniformity of climate have been established and controlled by solar heat alone. Hence during this period earth heat was the controlling source.

This era merged gradually into the Mesozoic era of tropical heat, during which the forms of life developed into higher types, and their range of distribution demonstrates the still perfect uniformity of. climate. One peculiar and significant fact is recognizable in comparing the land forms with the marine forms of life. The former developed types more suitable to tropical climates, while the latter held more tenaciously to the torrid types, thus proving the more rapid loss of heat by the continents.

The fossil life of the Cenozoic era corroborates to a remarkable degree the still perfect uniformity of climate. Throughout Greenland, Iceland, Lapland and Spitzbergen, as well as throughout present temperate and tropical zones, a perfectly uniform and temperate climate existed. The flora and fauna of the lower Mississippi valley flourished in those localities in which, during the Palæozoic era, only gigantic *Ferns*, *Lycopods*, *Calamites* and corresponding plant and animal life could be found, and where now only a stunted Arctic life can exist.

The palæontological evidence of the Mesozoic and Cenozoic eras is equally convincing as to the perfectly uniform tropical climate of the one and temperate and later frigid climates of the other.

During the latter part of the Tertiary and early Quaternary periods identical types of life existed in all parts of Europe, Asia and America and a uniformly temperate climate prevailed over the whole northern hemisphere entirely at variance with the extreme range of temperatures now embraced in that half of the globe.

The control of the waning earth heat was simply dying out, and had reached that stage in which it was no longer able to maintain the high temperatures of previous eras.*

The evidence that the high specific heat of water held the last available remnant of earth heat, and thus perpetuated its control of climates, is beyond dispute, as presented by the conditions culminating in the Ice Age.

Whatever may be the doubts as to the actual date of the Ice Age, there is no disputing the fact that the evidences establishing the culmination of that Age are found *above* or since the Tertiary, and *below* or before the Modern Era.

---

* The author is aware that this statement is at variance with the opinion of many Geologists of high repute, as may be seen from the following quotations: "It is evident that the idea of connecting the phenomena of the internal heat of the globe with terrestrial climates, whether of the present or of past geological ages, must be entirely abandoned, as it has been, by most writers on this subject. The hypothesis cannot be allowed to stand as even one of the possible theories of climatic change."— *The Climatic Changes of later Geological Times*. Whitney, p. 261.

" The first theory brought forward to account for glaciation was that the earth, having been originally in a fiery state, had in cooling passed from a condition of universal warmth to a more and more frigid state, until the present conditions were attained. This is the least tenable of all theories, for it neglected the now evident fact that there had been changes from cold to warmth and back again to cold. However, as it was invented before the existence of glacial periods was suspected, it long commanded a general assent, and was the opinion that held the ground until near the middle of this century."—*Glaciers*. Shaler & Davis, p. 70.

The physicists who have held that earth heat was a cause of the Ice Age are Prof. E. Frankland, F. R. S., Prof. A. Woeikof and Sartorius von Walterhausen. Not one of the three, however, seems to have had a clear conception of all the facts and conditions although their views were in the main sound.

The author hopes to extend the views held by these writers and to show that the whole range of climates as recorded by fossil and existing life is capable of correct interpretation, in accordance with known laws, and without the intervention of suppositions and assumptions. And moreover, to base his deductions upon a general plan applicable to any planet and capable of explaining conditions prevalent upon other planets, notably upon Jupiter and Mars.

Between these two periods there is abundant evidence from every climate, from every zone of present life, that the continents were glaciated.

Europe and Asia, North and South America, Africa and Australia,* all present glacial striæ, boulder deposits, and other marked evidences of glaciation at the same period, just antedating the Modern Era, or during the Quaternary period.

When we examine the evidence found in one of the present climatic zones, this change of climate from an ultra-torrid successively to a torrid, tropical, temperate, and, lastly frigid character, is not only very marked, but is everywhere the same.

---

* To those interested in a verification of this very wide distribution of glaciation, the following short list of authorities is recommended:

Asia.—The Great Ice Age (Giekie); Note on the Glaciation of parts of the Valleys of Jhelan and Scind Rivers, in the Himalaya Mountains of Kashmere, Lat. 34° N. (Capt. A. W. Sleff), F. G. S.; Jour. Geol. Soc., London, vol. xlvi, p. 66; Mem. Geol. Survey of India, vol. xxii, also vol. xiv; Record Geol. Survey of India, Nov., 1880; Jour. Asiatic Society, Bengal, xxxvi, p. 113; Brit. Association Report, 1880; Text Book of Geology, A. Giekie, LL. D., etc., p. 911.

Europe.—The European Glacial Literature is too extensive to mention.

America.—The Ice Age in North America (Wright); U. S. Geological Reports; State Geological Reports of Ohio, Kentucky, Pennsylvania, New York, Minnesota, Nebraska, Colorado, Illinois, etc.; Virginia, Am. Jour. Sci., vol. vi, p. 371; California, Am. Jour. Sci., vol. iii. p. 325, vol. x, p. 26.

South America.—Geological Sketches, Agassiz, p. 154, *et seq.*; Geol. and Physical Geog. of Brazil (Prof. Ch. Fred. Hartt), pp. 22, 28-9, 469-70, 490, 558.

Africa.—Geol. of South Africa (Stow); Quart. Journ. Geol. Soc., London, vols. xvii and xviii.

Australia and New Zealand.—Climate and Time (Croll), p. 295; Am. Jour. Sci., vol. 32, third series, p. 224; Proc. Linnæan Soc., N. S. W., May, 1886; Prestwich's Geol., vol. ii, p. 467; Rep. Brit. Assn., 1881, p. 742.

"The shrunken or vanished ice of mountain ranges is indeed equally characteristic of the Himalaya, the Lebanon, the Alps, the Scandinavian chain, the great chains of North and South America, and of other minor ranges and clusters of mountains."—Ramsay, Quart. Jour. Geol. Soc., 1862, p. 204.

Upon the establishment of solar heat in the control of surface temperatures, we find the isotherms entirely at variance with those ante-dating the Ice Age. We find also strong corroboration in the lines of retreat of the continental ice caps. These lines are sensibly parallel with the isotherms established by solar heat, proving that solar heat was the cause of the disappearance of glacial conditions.

These facts distinctly prove the totally different source and distribution of heat before and since the Ice Age, and that upon the inauguration of climates controlled by solar energy, an obliteration of the conditions left upon the dying out of earth heat set in. Along that zone most exposed to solar energy conditions and life corresponding to the tropical conditions of Cenozoic times have been established; along those zones moderately exposed to solar energy the newly established conditions are analogous to the universally temperate climate of the latter Tertiary and early Quaternary periods; whilst in those zones least exposed to solar energy a removal of glacial conditions is yet in progress.*

Wherever fossil life has been developed the order of climates, as thus recorded has been: First, torrid; second, tropical; third, temperate; fourth, frigid; and fifth, the life appropriate to the zone of solar climate— irrespective of that existing previous to the Ice Age. The same order is true for any portion of either temperate zone; under the equator the order of climates

---

* Many geologists are misled by the greater modification of tropical drift by sub-aerial agencies. Having been longer exposed to such agencies, greater modifications are to be expected. The apparent improbability of tropical glaciation seems to deter many scientists from believing that such glaciation could ever have occurred, yet the same scientists will accept the fact that fossil life establishes the existence of tropical or even torrid conditions within the polar circles during past ages.

has been the same, except a return to tropical conditions and life.*

In the North frigid zone this same order of climates has been found, except that there has been no change from the conditions left upon the dying out of earth heat; in other words, solar energy has not removed glacial cold in those regions least exposed to its action.

The removal of glacial conditions has been less in the Antartic than in the Arctic regions, partly from causes pointed out by Maury, and more fully treated by Dr. Croll. This removal has also been subjected to variations due to the mild astronomical influences ascribed by Adhémar, Croll, Ball, Drayson and others, as sufficient to produce glaciation.

These astronomical causes undoubtedly must have produced slight secular variations in the relative exposures of the two hemispheres to solar heat—but they have not been demonstrated to be of sufficient influence to produce glaciation, and in no way could they sensibly affect climates prior to the establishment of the control of solar heat. (See page 41.)

The distribution of heat, prior to the Ice Age, as recorded by fossil life, being entirely at variance with that now found, and being entirely independent of proximity to, or distance from, the equator, distinctly proves that climates were established and maintained independently of solar heat, and hence belong to the only other source, viz., earth heat.

It is also evident that under no possible conditions could solar energy maintain a torrid, tropical, temperate and lastly glacial climate over the whole range of the present zones of climates, and that this uniform distri-

---

*See Geological Sketches, Agassiz, p. 154, *et seq*. Also Physical Geography and Geology of Brazil, Prof. Ch. Fred. Hartt, pp. 22, 28, 29, 217, 469, 470.

bution of heat prior to and during the glaciation of the globe was due to an evenly distributed supply from a constantly and uniformly decreasing source.

Moreover, the wide distribution of glaciation over the present temperate and torrid zones is a distinct proof of the exclusion of solar heat from these regions during glaciation. Under no possible circumstances could temperate North America, Europe and Asia and tropical South America have been glaciated unless these regions were shut out during glaciation from that solar energy, which when admitted has removed glacial conditions.

Glacial dispersion followed one of two general laws: First, the great centers or belts from which dispersion took place in apparent disregard of the slope of the ground were areas most exposed to cyclonic activity and resulting precipitation. Second, minor centers of dispersion (or local glacial dispersion) were elevated lands, subjected to uniform precipitation.*

Since glacial conditions in the northern hemisphere were removed from southerly towards northerly latitudes, the gradients were increased southerly and decreased northerly from lines of maximum glaciation. Glacial transportation was likewise modified. The reverse of these directions prevailed in the southern hemisphere.

Glaciated areas have been partly relieved of their loads of ice at rates and times proportional to solar exposure, and upon lines parallel with present mean annual isotherms. Wherever remnants of the continental ice sheets of the Ice Age yet rest, this retreat is still in progress from the same cause.

The ascription of great elevations above sea level dur-

---

* See the author's views in *Physical and Geological Traces of Permanent Cyclone Belts.* Trans. Technical Society of the Pacific Coast, Vol. VIII, No. 1, June, 1891.

ing the Ice Age is natural, and such apparent greater elevation is due to *two* causes during this period, whilst due to only *one* cause during previous eras. As the surface of the earth became subjected to a temperature of 31° Far. under the oceans, and a corresponding temperature under the continental ice caps, contraction and consequent elevation were continued as before; and as snow was piled up upon the continents, water was withdrawn from the oceans; for each million square miles of continental ice cap three hundred feet thick a corresponding three million square miles of ocean was lowered one hundreed feet. The continental ice caps already approximately known were too vast not to have lowered the sea level to a marked degree.

The apparent general depression after the Ice Age is as natural. By the melting of the greater portion of the ice caps, and the evaporation of vast inland seas, the sea was approximately restored to the level existing prior to the Ice Age, thus causing an apparent sinking of the land.

The great difference between climatic conditions prior to and since the Ice Age is very marked around inland seas and basins without drainage. Lake Bonneville and Lake Lahontan, in the United States, and the greater area once occupied by the Caspian and other seas, evidence the superior dampness and rainfall antedating the Ice Age. During the control of earth heat the oceans were heated to their bottoms, and furnished moisture enough to keep these great depressions full of water and to support a dense life upon now desert areas. The dry air of the modern era has not only absorbed the water in these vast lakes, and restored it to the oceans, but vaster areas have been converted into deserts by the unequal distribution of heat and moisture under solar control.

3

PALÆOZOIC GLACIATION.

" Glacial Periods."

It is probable and may be regarded as a fact that upon certain of the oldest and highest mountains, glaciation was inaugurated during the Palæozoic Era, to slowly disappear by the gradual setting free of earth heat by vast fractures of the crust or to remain as local glaciation until the Ice Age. Isolated glacial deposits of this nature which were independent of solar exposure readily account for the early "Glacial Periods," which were evidently local phenomena antedating the Ice Age. It is neither logical nor reasonable to interpret the finding of evidences of early local glaciation into a Glacial Period, for local glaciations are found now in the Alps and upon certain peaks of the Sierra Nevada, and even in the torrid zone, but they by no means establish the present existence of a Glacial Period.

Evidences of glaciations antedating the Ice Age are wholly of a mechanical nature—namely, the transportation of boulders, striæ, etc. No corroborative evidence of fossil life of Arctic habits has been found. This is particularly the case of marine fauna and flora, which may be held as the only indisputable evidence of an Ice Age.

Granting that the evidences found be sufficient to establish Palæozoic glaciations, the absence of fossils of an Arctic type proves such glaciations to have been local and possibly of short duration, for had such glaciation been general and of long duration both plant and animal life would have been modified into temperate and Arctic types, as occurred later when general glaciation ensued.

It is apparent that the isotherm 32° Far. could have shrunk for a short period to the tops of mountains and that glaciers could have formed and coursed their way into a subtropical growth below; and that these conditions would be removed by the setting free of earth heat with the consequent rise in temperatures.

These changes followed too closely or were too limited in area to permit the evolution of forms of continental life adapted to temperate and Arctic conditions.

Palæozoic glaciations in no way conflict with the demonstration herein given—they are really coroborative of the other facts advanced to prove that prior to the Ice Age solar heat was shut out from the surface. For the evidences of Palæozoic glaciation occur in temperate and tropical latitudes adjacent to fossil life indicative of high temperatures. Early glaciations were dependent only upon elevation, and latitude did not influence their occurrence in any way whatever, and whether in Norway or India these glacial conditions were coexistent with tropical life at a lower elevation and equally independent of latitude.

When the crust became too thick and non-conducting to yield a sufficient supply of heat to hold mean temperatures at a higher degree of heat than 32° Far. this isotherm shrunk to the surface only to be removed by solar heat. Since the position of this isotherm was independent of latitude its intersections with the surface depended only upon elevation, and as the continents lost their heat more rapidly than oceans, the latter were the last to fall to 32° Far.

As in previous eras torrid, tropical and temperate life had existed in Spitzbergen, France and Brazil independent of the latitudes of these countries, so too were glacial conditions equally independent of latitudes. But in the removal of glacial conditions the isotherms of solar climates were necessarily followed.

Thus the Ice Age marks the date at which the climates of the globe passed from the control of earth heat to that of solar heat.   The great specific heat of water retained in the oceans, the energy necessary to maintain the cloud shield shutting out solar heat until both land and ocean areas could be equally cooled and contracted, thus ensuring the maximum degree of thickness and stability to the crust.   The mysteries of geological climates, interpreted by known laws, applicable alike to all members of the solar system, develop thus into a system beautiful in its simplicity.

Once realize that the surface temperatures of the globe were at one era in the past too high, by reason of *internal heat*, to permit water to remain upon the surface, and the peculiar properties possessed by the various forms of water and their relations to heat and cold, and follow out these facts to their natural and logical conclusion, and the whole mystery of geological climates clears up and becomes simple.

MATHEMATICAL CALCULATIONS AS TO THE DURATION OF EARTH HEAT.

It would not be proper to make the foregoing interpretation of the cause of geological climates without briefly referring to the various mathematical calculations which have been made by high authorities, and which reach conclusions materially differing from the deductions herein presented.

The arguments of Sir Wm. Thomson[*] and others, to the effect that internal or earth heat could not have affected the climates of the globe, by reason of the non-

* Phil. Mag. (4), Vol. XXV, pp. 1-14. Trans. Royal Soc. Edin., Vol. XXIII. Influence of the Earth's Secular Heat upon Climates. Hopkins, Jour. Geol. Soc., Vol. VIII.

conductivity of a comparatively slight thickness of crust, are not conclusive. These arguments are based upon:

1. An erroneous assumption of the manner in which heat is lost by a planet, upon which there exists an atmosphere and a fluid possessing the physical properties of water.

2. No account is taken in these calculations of the heat set free by denudations, etc.

3. The conservative action of active exterior sources is not considered.

1. To assume that in a molten or nearly molten planet heat was lost by direct radiation from the heated surface, is to assume a mode of loss that could not possibly occur with the constitution of our planet, nor with one possessing a constitution generally similar.

At the period assumed as the starting point of these calculations the earth's crust was just forming from the molten state. At this period, which has undoubtedly existed, all uncombined water must have been evaporated, and must have existed as an enshrouding cloud, shutting out solar heat and *shutting in earth heat.* Our planet at this period must have presented an appearance similar to that now presented by Jupiter, whose available internal heat has evidently not yet been exhausted, and upon whose surface evaporation must be kept up by internal heat.

The loss of internal heat by a globe constituted as our planet, must proceed, not by the radiation and loss of heat directly into space, but by the performance of work in the expansion of water to vapor, the exposure of the upper or cold surface of the partly condensed vapor to loss of heat by radiation into space.

The existence of a non-transcalent cloud shield is geologically recorded in most unmistakable terms, as

previously expláined, by the maintenance of eras of
tropical, temperate and frigid climates from pole to pole—
irrespective of latitude; by the glaciation of areas over
which solar energy, when not thus shut out, was capable
of removing glacial conditions and establishing much
warmer climates; also by the contrast of geological
climates with solar climates, one independent of, and
the other mainly dependent upon, latitude.

Thus the loss of heat by the crust must have proceeded with great slowness; and the crust in thus cooling was, by the laws of cooling solids, made as thick as
possible.

2. The non-conductivity of this cooling crust was a
cause of the long, instead of a cause of short duration of
the internal heat, for when too thick to yield up its heat
by conductivity, additional increments were but slowly
set free by denudations, faults and fractures. The volume
of heat thus set free may be partly grasped when it is
considered that no portion of the crust can be reached
that is not built up of denuded materials. Heat im-
prisoned by a non-conducting crust is more certain of
liberation by denudation than if the crust were composed of strata having the conductivity of beaten silver.

The assumption that the low conductivity of the crust
was a cause of the short duration of earth heat as a controlling factor is exactly contrary to the actual tendency
of such low conductivity.

3. From the cold outer surface of the cloud envelop
heat would radiate much more slowly than from the
more highly heated surface beneath. Indeed, there
is every reason to assume that this upper surface may
have been partly composed of fine crystals of ice,
as cirrus clouds may now be. Upon this upper
surface, whatever may have been its condition, was

received every thermal unit of heat reaching our globe from exterior sources from its development to the culmination of the Ice Age. What calculation has considered this single factor? which, for aught we know, may have been but little less than the original available store. To what consideration is any discussion which omits this factor entitled?

Having properly assumed a temperature which would necessitate the evaporation of all uncombined water and its suspension above the heated surface, a scientist should follow the results to their legitimate and logical conclusion, and not neglect the existence of a condition necessarily coexistent with those assumed.

The removal of the enshrouding clouds need not be assumed; such removal was blazed upon the globe in broad zones of climate and life which only solar energy could maintain. These lines are as distinctly different from those written by earth heat as daylight is from darkness.

When this removal did take place the fact was graven upon our planet by the melting of the massive glaciers deposited during and before such removal, and by the establishment of the existing conditions.

All calculations and discussions omitting these three factors must reach illogical and erroneous conclusions. The omission of a single one would be fatal, and entitle the result to no farther consideration, and justifies the cynical view that " There is something fascinating about science. One gets such wholesale returns of conjecture out of such trifling investment in facts."

ASTRONOMICAL CAUSES AND THEIR INFLUENCE.

We will now consider the effect of astronomical causes. Dr. Croll has elaborately discussed the variations in solar exposure to which the two hemispheres of our planet have been and are subjected.*

The distinguished Astronomer Royal of the Dublin Observatory, Dr. Ball, has shown that 63% of the gross solar energy received by either hemisphere reaches it during summer and the remaining 37% during winter.† High authorities, both before and since these publications, have discussed various phases of these influences, as well as offered remarkable and unverifiable hypotheses regarding the temperature of space, solar energy, and the heat absorptive power of a solar envelope.‡ It is not necessary to attempt a discussion or elaboration of these views. Should the interpretation herein rendered be correct, it follows that variations in the distance from or degree of solar energy could not have directly affected the surface temperatures of the globe prior to the culmination of the Ice Age, and that only since that age could these slight variations have acted, except in a conservative way. It is unquestionable that for many years past the temperature of the northern hemisphere has risen more rapidly than the southern. This condition is proved not only by correct deductions from actual conditions and laws, but by observation. This is also recorded geologically by the greater removal of glacial conditions in the northern hemisphere—although in both this removal is yet in progress.

In a globe wrapped in a mass of vapor by reason of

---

* Climate and Time, Climate and Cosmology.

† The Cause of an Ice Age, chapters 5 and 6.

‡ Nature, May 1891; S. E. Bishop.

evaporation maintained at the surface by its own heat and condensed upon the outer surface of the spheroidal cloud envelope, it is immaterial so far as surface temperatures are concerned, to what degree of outside heat it may be subjected. The only possible effects of variations in the distance from, or intensity of the exterior heat source being to influence the duration of the interior supply and the distance therefrom at which cloud condensation takes place.

In a globe thus enshrouded the same order of surface temperatures would follow, whether revolving in the orbit of Venus or that of Neptune—the actual influences being the greater rate of loss in the remoter position, the more rapid succession of geological climates, and the greater time necessary for the removal of glacial conditions, and for the establishment of solar climatic control. Could the earth have been removed during the Archæan Age to the orbit of Jupiter without disturbing other conditions, no change could have occurred in the order of succeeding geological climates prior to the Ice Age. The rate of receipt of solar energy would have been in the ratio of $(5.2)^2:1$; and the actual retardation of loss would have been in this ratio, as also the rate of establishment of solar climatic control; the crust would have cooled quicker, and therefore have been thinner and less stable than at present.

The observed movements in the cloud envelope of Jupiter present phenomena warranting the belief that his atmosphere is non-transcalent.* In this particular it resembles the clouded atmosphere of the earth; and indicates a condition analogous to that of the earth in pre-glacial ages. The smaller planets by reason of their lesser masses have lost their available resident heat.

---

* Circulation of the Atmosphere of Planets.—Trans. Technical Soc. of the Pac. Coast, Vol. IX, No. 5; pp. 136-143.

Their atmospheres have become cleared and are both translucent and transcalent. Their surfaces can be observed, and their volumes and densities calculated with a reasonable degree of exactness. In the cases of the larger planets observations are confined to the surface of their spheroidal cloud envelopes, and hence to these planets are ascribed volumes and densities varying abnormally from those whose actual volumes can be measured. The satellites of Jupiter possess much greater densities than that ascribed to the great planet—were it possible to measure the actual volume of his enshrouded mass this apparent anomaly would be in whole or in part removed.

Not knowing the surface temperatures, the exact composition of the atmospheres, nor the dimensions of the planetary masses, the distances to which the cloud envelopes of Jupiter, Saturn, Uranus and Neptune may be expanded, are yet matters of conjecture. Whatever is known of these planets corroborates the interpretation herein rendered of the record of the geological climates of the earth.*

THE ESTABLISHMENT OF SOLAR CLIMATES.

At the culmination of the Ice Age the snow line was much lower than at present, and elevated lands† at all latitudes were deeply glaciated; the seas were intensely cold. It is evident that since the culmination of the Ice Age and in the establishment of the present climates there has been a great rise in temperatures in the tropi-

---

* For a further discussion of the conditions prevalent upon Jupiter see *Circulation of the Atmosphere of Planets*, previously quoted.

† Except hot or warm lava covered areas, and the protected or "unglaciated areas" to the eastward of such lava overflows. (See page 44 and note (†) page 14.)

cal, temperate and sub-frigid zones. There is also indisputable evidence that this rise in temperature is yet in progress. This accession of heat must therefore be accounted for by the correct application of laws and forces now-acting, and it is not necessary to go outside of these known laws and forces to render a correct interpretation of the establishment and maintenance of the zones of climate now existing.

It will be observed that when the oceans were exhausted of their heat and the lands deeply glaciated, the crust was shrunk in upon the interior mass by being uniformly chilled down to the lowest temperature to which a planet, upon which water and an atmosphere exist, can be subjected. The atmosphere was then cleared of clouds and heat rays from exterior sources permitted to reach the planetary surface. At once these rays began to be changed into dark heat rays, particularly those from water, and the trapping of heat ensued; from this date a general rise in temperatures must follow from the accession of heat from exterior sources, until checked within the moderate limits hereafter outlined.

The trapping process thus inaugurated is independent of the actual amount of heat received whether from solar or stellar sources.

Were it possible for the now pent-up internal heat to raise the temperature of the oceans, the crust at the bottom of the oceans, and under the polar ice caps to a mean temperature of say 68 degrees Far., the accession of heat from exterior sources would be shut off, as in early Quaternary times, by dense clouds; the exterior would be again shrunk by glacial conditions, the air cleared as before and heat from exterior sources in whatever amounts it then reached the surface would be trapped as succeeded the Ice Age.

This action must in turn take place upon any planet upon which water and an atmosphere resembling ours exist. The rate at which a planet acquires heat from exterior sources is dependent upon the power of its atmosphere to trap heat; very slight variations in the atmospheric constituents producing great variations in heat trapping power.

Orbital distance being only a function of the amount received and not of the trapping process, this rise in temperature is as certain to follow in one position as another.

By thus being subjected to the maximum shrinking-strains the weakest portions of the crust were ruptured. The lava ejected from these ruptures was spread out over the weak areas in successive layers of a few dozen feet in thickness until the added strength reached that belonging to thousands of feet of solid rock.*

To digress a moment—

These lava overflows evidently performed another important function. The heat set free by each successive layer could not have been lost by radiation into space, for the enshrouding clouds had not yet been removed. The air and clouds caught this heat and bearing it eastwardly in their general course caused warm rains instead of snow to be precipitated upon the adjacent region. In this way the "unglaciated area" escaped glaciation; in

---

* The Columbian Lava Plains of North America aggregate some 150,000 square miles; the Deccan Lava Plains of India cover an almost unbroken plain 200,000 square miles in area. No Geologists ascribe these lava overflows to an earlier date than the Tertiary; the author could find no reason to assign the Columbian Lava Plain to so early a period, and strong reasons to assign the continuance of the flow to the later Quaternary; of the Deccan Plain he is unable to speak. (See Trans. Geological Society of Australasia, vol. i, part vi, p. 162, note. Also Mining and Scientific Press, Feb. 6th, 1892.

this area are the "bad lands" of Dakota, whose topography distinctly shows that sub-aerial denudation, and not glacial ice formed the controlling features. In this area are the great deposits of tertiary fossil life, in perfect form—uncrushed by the mighty tread of the glaciers which surrounded them on all sides, except to the west. From areas such as these went forth the life that survived the glacial winter.

That the isotherm marked by glacial ice is yet slowly retreating upward is recorded not only by tradition and history but geologically and physically, as observed by every scientist who has studied existing glaciers.*

This retreat is a positive proof of either a decrease in precipitation on the tributary areas, a rise in temperature, or both of these agencies acting conjointly. There is no evidence to show that a decrease in precipitation† is synchronously taking place over the sub-frigid, temperate and tropical regions of both hemispheres, as is the retreat of glaciers; and there are positive and active causes in force which have affected, and are yet affecting an increase in temperature. We must therefore conclude that this rise in the isotherm marking glacial ice

---

* *Climatic changes indicated by Glaciers.*
Prof. I. C. Russell, Am. Geologist, May, 1892, vol. ix, No. 5. In addition to the very extensive list of authorities there quoted by Prof. Russell, see also Report of The British Ass'n. 1881, p. 742.
Life of Agassiz, Vol. II, pp. 717 to 729 and pp. 743 to 747.

† At the culmination of the Ice Age evaporation reached its minimum, and hence precipitation was also at a minimum. Since that Age evaporation has slowly increased; the amount of moisture in the atmosphere being dependant upon its temperature, this amount has also increased. The aggregate amount of evaporation and the aggregate amount of precipitation is slowly increasing, and has the moderate limit fixed by natural laws for increase of mean temperature. Mars appears to have progressed further in this mean condition than the earth. The smaller mass partly accounts for this.

is due primarily, if not entirely, to an accession of heat.

It has been demonstrated that at the culmination of the Ice Age, much colder conditions existed than at present. It now remains to explain the conditions acting to bring about existing climates. Upon the exhaustion of the last available remnant of earth heat—left in the oceans by reason of the high specific heat of water—the supply of vapor maintaining the cloud envelop was shut off, and solar heat permitted to reach the planetary surface.

That direct solar rays are converted into obscure or dark heat rays by contact with the planetary surface, and that the atmosphere of our planet is more transcalent to the former than to the latter, has been fully demonstrated by Tyndall,* although slightly modified by Buff.†

However small may be the difference between the transcalency of the atmosphere to direct solar rays and to the dark rays into which the direct are converted, a gradual rise in temperature must follow. This rise must follow whether solar energy be constant or slowly decreasing, the rise being due not to the actual amount of heat received, but to the difference between the rate of receipt and the rate of loss.

The great increase of mean surface temperatures in equatorial, temperate and sub-tropical areas being due to this small but positive difference between the rates of receipt and loss; as has just been shown, this action is yet in progress.

These deductions are radically at variance with the

---

* Archives des Sciences, vol. v, p. 293. Proc. Royal Soc., vol. xiii, p. 160.

† Archives des Sciences, Berne, vol. lvii, p. 293, *et seq.*

opinion of high authorities on meteorology, as may be seen from the following quotation: "It is evident that our planet, considered as a whole, and on the average of many years, loses all the heat that it receives from the sun, but all the details of this process have not yet been worked out.*"

The author is unable to find any facts to sustain this view—all tend to refute it. The trapping of heat by vapors and gases of the atmosphere—the gradual retreat of glaciers in both hemispheres—and the vast rise in temperatures since the culmination of the Ice Age—all conclusively tend to corroborate the deductions just reached—namely, that the mean surface temperatures of the globe have been and are yet rising from the trapping of heat.

It does not follow that this rise has an indefinite or excessive limit, as the oceans become warmer they are cooled by giving off more vapor. This vapor, when partly condensed into clouds, intercepts solar heat in the upper atmosphere, and the intense white of the upper surface of clouds reflects more heat into space than the darker planetary surface beneath.†

The vast store of cold in the continental ice sheets has been greatly exhausted; there yet remains the vaster store in the ice cold depths of the oceans, the conservative influence of which cannot be estimated; for besides the difficulties of heating water from the surface down-

---

*Dr. Cleveland Abbe, U. S. Meteorological Bureau. Am. Jour. of Science, May, 1892, vol. xliii, p. 364.

† The albedo of Jupiter is 0.62; that of Mars, 0.26, of the moon 0.174. It will be observed that the planets distinctly shrouded in clouds have high reflective powers; those planets and satellites not shrouded have very low powers. Venus, in this respect, seems to have a partially obscured atmosphere, her albedo being 0.50.

wards, there yet remains the cooling effect of surface evaporation. There is thus presented the extreme slowness of the methods by which vast changes are wrought. Here are agencies whose results are so slight as not to be detected by thermometric methods—yet recording their effects in grand eras of climates throughout the earth.

The planet Mars is particularly interesting, having a mass less than one-ninth ($\frac{1}{9.4}$) that of the earth. His loss of internal heat occurred ages before that of the earth; therefore, Mars has been a heat-gathering body longer than the earth, and enjoys a milder general temperature,* although that planet receives less than half the heat and light received by the earth. Jupiter is in a condition which our geological history proves the earth to have passed through; Mars is in a condition towards which the earth is gradually tending.†

It is now a simple matter to trace the steps by which glacial conditions were removed and zones of climate established.

Solar energy first established its control in that zone most exposed to its power—namely, the torrid zone. From this zone glacial conditions were first removed, and this removal continued north and south upon lines parallel with present isotherms.

In considering the astronomical causes, and the physical results thereby brought about, it was argued that these causes tended to heat the northern hemisphere more rapidly than the southern. Dr. Croll and other physicists, have so fully discussed this question that there remains but little to be added.

---

* General Astronomy.—Young; p. 337.

† Venus presents a condition which suggests that she may be partly shrouded in clouds, shutting out solar heat, just as the thermal equator of the earth is thus partly protected by the equatorial cloud ring.

The prime reason, however, seems to have been omitted, which is simply this, the northern hemisphere, containing so large a predominance of land area, was more easily warmed than the southern. This unequal heating once inaugurated would establish currents both of air and water tending to perpetuate this action, reinforced as it is by geographical and cosmical agencies.

When, by this gradual accession of heat, conditions and temperatures resembling those existing prior to the Ice Age, were re-established, we find these new conditions restricted to latitudinal belts sensibly parallel with the equator, but modified by elevation and ocean currents; whereas the corresponding pre-glacial climates were independent of latitude.

By the trapping of solar heat a gradual rise in temperature was inaugurated at that period, when by the exhaustion of the earth heat, left in the oceans, the enshrouding clouds were removed. Then, and not until then, do we find the removal of conditions shutting out solar heat written in zones of life belting the earth. In these new zones of climate there have been developed higher, nobler types of life, and with the birth of the seasons there was ushered in upon the earth that Light which is developing Psychozoic Life.